Heavy 24 cm Cannon Development and Action 1916-1945

Early October 1941: A heavy 24 cm cannon firing. Shell after shell leaves the barrel with the greatest possible rapidity. While one cannoneer seizes the still-smoking cartridge case, his comrades heave the next shell, weighing almost 200 kg, onto the loading tray.

Wolfgang Fleischer

Schiffer Military/Aviation History
Atglen, PA

Dibliography and Data Credits

—Federal Archives, Koblenz, Potsdam Branch
   WF 10/2483. WF 10/2498;
—H.Dv. 481/158, Merkblatt für die Munition der schweren
   24-cm Kanone (t), Berlin 1941;
—Beinhauer, Artillerie im Osten, Berlin 1943;
—Hahn, Waffen und Geheimwaffen des deutschen Heeres
   1933-1945, Koblenz 1986 and 1987;
—Handsche & Clauss, 1941—Vom Cap Gris Nez bis
Leningrad, place unknown, 1941;
—Hogg, Deutsche Artilleriewaffen im Zweiten Weltkrieg;
   Stuttgart 1978;
—Schmitz & Thies, Die Truppenkennzeichen 1939-1945,
   Vol. 1, Osnabrück 1987;

Periodicals

—Artilleristische Rundschau, various issues;
—Deutsche Wehr, No. 23/24, 1936;
—Der Frontsoldat, No. 5, 1955;
—Militärtechnik, 1971;
—Waffenhefte des Heeres—Die Artillerie—Die Wehrmacht,
No. 22, 1941;
—Information from Mr. Gerhard Fritzsche of Radeberg and
Mr. Helmut Szajny of Zwickau.

Photo Credits

BA Koblenz (1), Fleischer (86), Schüler (2), Thiede (1)

A generator truck pulls an electric chassis with a gun barrel off a railroad car.

Translated from the German by Ed Force

Printed in China.
ISBN: 0-7643-0569-7

This book was originally published under the title,
*Waffen Arsenal-Schwere 24-cm Kanone Entwicklung und Einsätze
1916-1945*
by Podzun-Pallas Verlag.

We are interested in hearing from authors with book ideas on related
topics.

Acknowledgments

The author thanks Frau Elke Schüler (drawings), Mr.
Gerhard Thiese (photographic work), and Dr. Lachmann and
Dirk Hensel for their assistance.

Published by Schiffer Publishing Ltd.
4880 Lower Valley Road
Atglen, PA 19310
Phone: (610) 593-1777
FAX: (610) 593-2002
E-mail: Schifferbk@aol.com
Please write for a free catalog.
This book may be purchased from the publisher.
Please include $3.95 postage.
Try your bookstore first.

Development and Action History, 1916-1940

The heavy 24 cm cannon was used by three European armies in the course of barely thirty years, and in this period it was always in use in certain geographical areas, which find special mention in the military histories written in later years. Its history begins in Austria-Hungary.

The Danube monarchy, thanks to the highly developed heavy industry in Bohemia and Moravia, already had a fine array of heavy and extra-heavy guns of modern type at the beginning of World War I. They included the 30 cm M.11 mortar and the 42 cm coast howitzer. During the war these were joined by:
— the 30.5 cm M.16 mortar,
— the 42 cm M.16 howitzer,
— the 42 cm M.17 autohowitzer,
— the 38 cm M.16 howitzer, and
— the 24 cm M.16 cannon.

Aside from the 42 cm coast howitzer, which was first used in 1915 as makeshift protection for railroad transport, the guns were motorized and thus very mobile.

In 1915 the Ministry of War contracted for two 38 cm howitzers. On January 5, 1916 the k.u.k. army command was shown the first finished gun at the Skoda works. The Ministry of War now set out to adapt the mount of the 38 cm M.16 howitzer for a 24 cm cannon as well. Within a short time, the cannon was developed, built and tested. It was introduced with the designation "24-cm-Kanone M.16". Of the nine guns that were ordered, only two more reached the troops by the end of the war.

The 24 cm M.16 cannon was a modern recoiling-barrel gun with box bedding and a built-in turntable. This made possible a traverse field of 360 degrees. The low mount guaranteed full shooting silence at all elevations and traverses. This fact was always mentioned especially in action reports.

The gun weighed 79 tons ready to fire. It was transported in four loads with weights between 30 and 38 tons. Demands were thus made of the transport vehicles that could be met only with designs unconventional under the circumstances of the times. It was a task for Dr. Porsche, who already ranked among the well-known designers at that time, and who in 1912, under contract to the k.u.k. Army Vehicle Department, developed the "Landwehrzug". Here the problem of transporting heavy payloads had been solved by spreading the burden among as many wheels as possible. The Landwehrzug consisted of a generator vehicle with a gasoline engine and generator and up to ten one-axle trailers, which had electric motors in their wheel hubs.

This principle was adopted for the "C-trains" of the 24 cm cannon and 42 cm howitzer, which were transported in four loads. In the normal type of transportation, the road train, an M.16 artillery generator vehicle moved an M.16 electric chassis with a one-gun load. The gasoline engine of the generator vehicle produced 150 to 165 HP at 1200 rpm and provided the current from the generator to drive a total of ten electric motors for the whole train (two electric motors in each Artillery Generator Auto M.16 and eight in the electric chassis).

This picture, taken in 1918, shows the gun in loading position. On the swinging loading tray, the 215-kilogram shell lies ready. Note the cover to keep the rain off the aiming apparatus on the left side.

Left: The 30.5 cm M.16 mortar in firing position.

Right: The mount of the 30.5 cm M.16 mortar before being lowered onto its bed. This method was also used for the 24 cm cannon.

Below: The 42 cm M.16 howitzer, with the gun mount attached to the bed and the barrel loaded on so that it can be pulled over the mount.

Originally rubber tires had been planned for the C-trains. The chronic shortage of rubber soon compelled the use of iron wheels. With that, the maximum possible speed sank from 16 to 12 kph. To protect the equipment, 10 kph was prescribed. The climbing ability was astonishing, being between 26 and 29%.

Another means of transportation was moving the gun loads on normal-gauge railroad lines. For this, the road wheels had to be removed. Short stretches of up to 50 kilometers were traversed this way. The top speed for this means of transportation was 27 kph, the maximum climbing ability 9%. Marches by rail over longer distances were carried out using normal transport trains.

With the 24 cm M.16 cannon, Austria-Hungary could use motorized heavy flat fire, to which Germany, for example, could offer nothing comparable. This lack soon became obvious on the western front. Railroad and railroad-protection guns in particular utilized this caliber, with all their weaknesses. Their use was linked to a productive railway network; the spread of firing positions was laborious and could also be spotted quickly by the rails to the positions.

The following table shows that the 24 cm cannon also had nothing to fear in terms of firing technology:

	Theodor Karl	Theodor Otto	M.16
Caliber (cm)	24	24	24
Barrel length (caliber)	L/40	L/40	L/40
Weight ready to fire (kg)	109,000	113,000	79,000
Shell weight (kg)	151	148	215
Maximum range (m)	26,600	18,700	26,300
Zero velocity (m/sec)	810	640	750

The Austrian cannon was not dependent on railway lines, which made the surprising formation of artillery focal points much easier. Depending on the type of terrain, readiness to fire could be achieved in 14 to 28 hours, while 2.5 to 6 hours were necessary for dismantling.

The military forces of the Czechoslovakian Republic took over extensive war materials as part of its inheritance from the fallen Dual Monarchy. Among them were the two 24 cm M.16 cannons, the numbers of which in the CSR were increased to six. In the twenties, the guns were modernized. The iron wheels were replaced by rubber tires. In addition, the Czechs introduced a lighter shell weighing only 185 kilograms. The maximum shot range was now 29,600 meters.

The further history of the 24 cm cannon is linked with a generally little-known but not uninteresting episode. In the thirties, the CSR was known to be one of the countries that exported goodly quantities of war materials. In 1938, guns, armored materials for fortresses, and other military equipment from the supplies of the Czech Ministry of War were offered for sale. The lists that went into circulation were examined with interest in Germany. The Chief of the Wehrmacht High Command, Generaloberst Keitel, came to the Reich's Foreign Ministry with the request to communicate with Prague about the offered war materials. The deal was ideal. Germany obtained eleven 15 cm cannons, 18 21 cm mortars, 17 30.5 cm mortars and the six 24 cm cannons. As to the asking price of the last, the War Ministry in Prague offered the following details and prices:

III.   Price per unit in RM

a. 6 units 24 cm M.16 cannon with complete equipment, including artillery generator auto M.16 and electric chassis M.16   602,000 RM

b. One replacement barrel   215,000 RM

c. Two school trains, consisting of one set of artillery generator auto M.16 and electric chassis M.16
        each 19,000 RM

d. 873 rounds of ammunition   1,290 RM

With the six 24 cm cannons, the motorized heavy artillery of the Army obtained a worthwhile strengthening. The quantities of Austrian and Czech weapons on July 1, 1940 included 42 heavy guns (21 to 42 cm calibers). There were also 79 guns of 21 to 30.5 cm calibers that had been made in Germany.

In April 1940, the Second Unit of the Army's Artillery Regiment 84 was equipped with the six 42 cm cannons.

The heavy 24 cm cannon in firing position, with its entire crew.

Below: A so-called C-train consisting of an Artillery Generator Auto M.16 and an Electric Chassis M.16 with its load, ready to run on railway lines. It can be seen that the tires have been removed.

This diagram shows all four gun loads for the 24 cm M.16 cannon, with the appropriate vehicles: 1. barrel wagon, weight 38,000 kg; 2. Mount wagon, weight 30,800 kg; 3. Bed wagon, left side, weight 36,600 kg; and 4. Bed wagon, right side, weight 37,600 kg.

Below: The Artillery Generator Auto M.16 with an empty Electric Chassis M.16.

Installation

Despite the great weight, the transition from marching to firing situation for the 24 cm cannon was relatively simple to carry out. The extra equipment required included four lifting jacks, several rollers and two guiding timbers with rails.

1. At first, an area of nearly 50 square meters of ground had to be raised to provide room for the bedding ditch (6.5 x 5.2 x 1.4 meters). To protect the bed, the ground was covered with wood. Other fundamental work was not required. (see picture)

The ground was slanted to one side so the timbers with the rails for the bed could be laid on it.

2. After that, the two bed autos were driven to the ditch in such a way that the two adjoining sides were opposite each other.

These pictures were taken during a practice assembly of the guns in the spring of 1940.

Below: In front of the electric chassis with the right half of the bed is the Artillery Generator Auto.

3. The bed halves were lifted with jacks and the electric chassis driven away; after that the rollers were put in place, under which the rails were laid. Now the bed halves could be lowered and brought together. Two centering spikes determined their exact situation. The united bed was rolled along the rails and over the bed ditch by the crew.

With a critical look, the battery chief observes the uniting of the bed halves.

4. Four jacks were put in place, the rollers removed, and the bed slowly lowered into the ditch with the help of the jacks.

5. Now the mount truck could be driven up and rolled over the bed on prepared approach rails. The mount had to come to a stop directly over the turntable. The mount was raised with jacks, the electric chassis driven away and the mount lowered and screwed in place. See also the next page.

*The mount is screwed onto the gun bed;
now the barrel can be installed.*

6. Finally, the barrel truck was driven up to the mount until
the barrel could be pulled into its cradle with lines. See also
the next page.

7. The sixteen-man gun crew, under the command of a sergeant, checked all the functions of the gun after it had been assembled.

Service of the II./Artillery Regiment 84, 1940 to 1945

Army Artillery Regiment 84 was established before World War II broke out, but had only the strength of a unit, the I./84, with three batteries of two German-made (K 3) 24 cm cannons each. The second unit was founded on April 20, 1940 and located in Zeitz. The three batteries of the II./84 were equipped with the six Czech 24 cm M.16 cannons. They now bore the official designation "heavy 24-cm Cannon M.16 (t)". The small t in parentheses was an indication of their country of origin. The two units of Army Artillery Regiment 84 never once experienced a joint action in the course of their existence.

After brief training, they took part in the 1940 campaign in Belgium and France. They were stationed before Liege and on the Somme. Because of the good road conditions, there were no particular transport problems. In terms of artillery, the guns proved themselves well. The usual artillery range-finding troops in the batteries were pointless. They each had one range-finding and shooting-in platoon (E.u.V.Zug).

In July 1940, parts of the unit were in Epinac les Mines in southern France. On July 16, 1940 orders were given by the highest military leadership that influenced the further history of the unit negatively.

In Directive 16 for preparation of the amphibious attack operation "Sealion", the OKH gave orders for coastal protection to be set up on the Netherlands, Belgian and French coasts. The securing of German ship movements in the Channel, as well as the fighting of enemy sea targets, made the presence of long-range heavy artillery on the Channel coast necessary. As early as July 15, 1940, the transfer of the batteries of the II./48 had begun. In four days' marches, the approximately 1200 kilometers to Audresselles between Boulogne and Calais were covered. North of the small town, Cap Gris Nez projects into the Channel. Its location offered ideal conditions for artillery securing of that sector of the coast.

In the night of July 21-22, 1940 the work of building the bedding ditch began, but it was complicated by the rocky ground. Some rock had to be blasted out. British reconnaissance aircraft had a good view of these activities. For anti-aircraft protection, Flak guns were put in position.

After thirty days of exhausting work, the firing positions were finished and the cannons temporarily protected by sandbag walls and camouflage. 120 km of telephone cable had been laid and several range-finding stations set up.

On August 17, 1940 Field Marshal von Brauchitsch visited the firing positions. This first prominent visitor was followed by others, including high German and foreign military officers, leaders of the Reichskriegerbund and the Nazi Party. The press and the UFA newsreel people were also there.

The troop symbol of the II./84, as could be seen on the unit's vehicles before Leningrad in 1941.

Other heavy artillery units gradually moved onto the cape. The Todt Organization began the job of building bunkers for the gun positions. On August 22 the cannons of the II./84 first opened fire. Their target was a British convoy in the Channel. In January 1941, a two-week training course was held at the batteries, ending with sharpshooting and inspection.

Early in March 1941 the batteries were on the move. The cannons were dismantled and sent by rail to the east, to Heilsberg in East Prussia. The preparations for the campaign against the USSR had begun. Hitler had already, on December 18, 1940, signed Directive 21 (Case "Barbarossa"). The II./84 had been assigned to the Army Group North as its only heavy artillery unit at that time. On June 17, 1941 the unit's guns moved into prepared firing positions on the border of the USSR, opposite Wirballen. In the morning twilight [p. 21]of June 22, the first shells were fired eastward by the guns. The unit followed the advancing front only on July 5. In the process of an exhausting march on bad roads, sometimes over swampy ground and unusable bridges, which demanded the utmost of the gun crews, they moved through Dünaburg and Ostrow to Pleskau. This was the end for the artillery generator autos. They could not stand the hardships, broke down and had to be replaced completely by 18-ton tractors.

Left: Even before the war began, the sister unit of the II./84, as the First Unit of Army Artillery Regiment 84, was established, with three batteries, each having two German-made 24 cm cannons (K.3). This picture shows a K.3 in firing position. The gun weighed 43,866 kilograms, the shell weighed 151.4 kg, and the shot range was 37,500 meters.

Right: The heavy 24 cm cannon in firing position. Note the low mount.

The II./84 went on to Luga and Krasnovardeisk. Their target was Leningrad, a city of three million, which was largely surrounded and besieged by German troops. In mid-September 1941, the unit's guns took positions before Kronstadt. Almost every day now, they fired on targets in Kronstadt and Leningrad, as well as the ships of the Baltic Red Banner Fleet which operated off the coast, among them the battleship "Mara".

In order to improve the effectiveness of the guns, their positions were changed again. The batteries were now very near the former imperial castle of Tsarskoe Selo (Peterhof). They remained in these positions until the beginning of 1944.

The battleships of the Baltic Red Banner Fleet remained a latent danger for the German troops. Fighting against them was complicated. For the 24 cm cannons of the II./84, a new shooting-in procedure was tested. For this, two range-finding positions were set up for each battery. In good weather, firing could also be directed by Fw.189 observation planes.

The observation posts of the artillery unit were the target of a Russian commando undertaking on the morning of October 5, 1941. Supported by the battleship "Oktyabrskaya Revolyuzia", naval ensigns landed near Snamenka. The attack was beaten off, partly in close combat.

In mid-October the thermometer went down everywhere. It became colder, and winter set in slowly. The gun crews were given orders to fire for which the maximum range of the cannons, some 29,875 meters, was no longer sufficient. Despite strong misgivings of the firemen, the shell cases were heated to 55 degrees Celsius, which increased their range to 31,600 meters.

Because of the increased pressure on the gun barrels, more and more of them broke down. By early October, the A gun of the 6th Battery shattered its barrel. Two weeks later, a gun of the 5th Battery broke down. The two batteries sent the remaining cannons to be overhauled and were transferred to Zeitz.

At the end of March 1942, the 5th Battery, with two overhauled heavy 24 cm (t) cannons, returned to its position before Leningrad; the 6th, equipped with three 17 cm cannons in mortar mounts, was likewise sent back to the Army Group North. The 4th Battery turned over its two 24 cm guns to be sent home for repairs in the spring of 1942, and was armed with three 19.4 cm GPF (f) cannons in Schneider 485 (f) self-propelled mounts. They saw action, among other places, at Staraya Russa.

The 5th Battery remained before Leningrad until early 1944. In January, a Russian offensive broke through the German ring around Leningrad. In ceaseless action, the

Another picture from the spring of 1940. During training, putting the cannons into position was often practiced. The smooth cooperation of the crew was of great importance in this complicated work.

supplies of ammunition on hand were used up. In the withdrawal, it was not possible to dismantle the mounts and beds; only the two gun barrels could be rescued. One was captured by the Red Army during the withdrawal.

Both of the guns that had been sent back by the 4th Battery in 1942 returned from the repair shop. They were taken over by the 5th Battery and remained with the Army Group North. They came into the possession of the 18th Army in the Courland pocket. As can be seen from reports, both guns were still in action in March 1945, and were with the troops along with 5244 shells. On May 9, 1945 explosive charges shattered the barrels of the last two heavy 24 cm (t) cannons.

Above: During the French campaign, an 18-ton tractor was already on hand as a reserve towing vehicle for each 24 cm cannon, so that it could step in if one of the four artillery generator autos, which were fairly old by then, broke down.
This picture was taken in June 1940 and shows a tractor with a bed load in tow.

Right: An M.16 artillery generator auto with a bed load, crossing the Somme on a makeshift bridge.

Left: Vehicles of the II./84 on the march from Epinac les Mines to Audresselles in June 1940. On the well-built French roads, transporting the loads weighing up to 38 tons was scarcely a problem.

Immediately after their arrival at Cap Gris Nez, the building of the bed ditches was begun. The rocky ground made the work much more difficult. Only after some time could the guns be put into position.

A picture of concentrated firepower: the guns of the 4th and 5th batteries in position.

Above: The guns of the II./84 at Cap Gris Nez were surrounded by a protective wall of sandbags and covered with a moveable camouflage screen. Open gun positions, like those on the cape, offered a number of advantages: Sea targets could be engaged at a higher rate of fire and with greater accuracy. The spotting of newly arriving sea targets took less time. But there were also disadvantages: The firing positions were visible from a long distance away and could be attacked effectively. An optimal solution was the use of bunkers as gun positions.

Below: In order to avoid damage to the equipment, longer trips were made by rail. In April 1941 the guns, after a long trip, arrived in Heilsberg, East Prussia. This picture shows the gun mount loaded onto a flatcar.

Going out to practice. The time in Heilsberg was used for the maintenance and care of the guns and equipment and the training of the gun crews. This picture was taken in Heilsberg in May 1941. In front is an 18-ton tractor towing the barrel load; in the background is an artillery generator auto with a bed load.

Below: June 22, 1941: At the first light of day, the guns of the II./84 start to fire on the Red Army positions on the other side of the Soviet border. Glaring balls of fire blind the observers behind the battery positions.

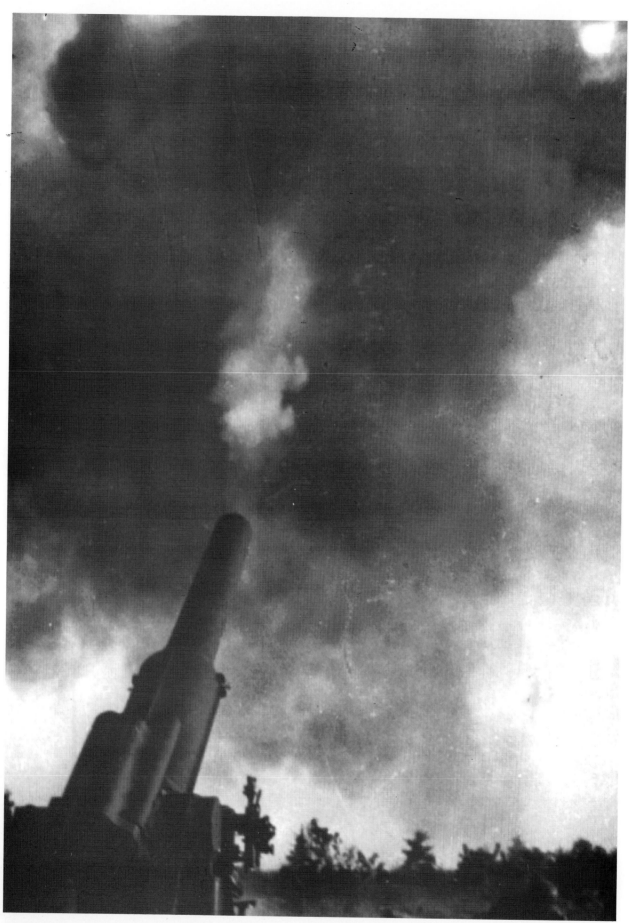

Fire! The tremendous air pressure, as observers of the events on the morning of June 22, 1941 reported, blew off the roof of a barn a goodly distance away from the firing position.

The barrel at its maximum elevation. In the picture, the loading tray can be seen at left, folded over.

New firing commands are sent to the batteries. Again and again, great tongues of flame come out of the gun barrels and the powder smoke darkens the skies on the morning of the fateful 22nd of June 1941.

The advance of the II./84 to the east took place from the beginning on bad roads and paths and was very difficult. One after another, all the artillery generator autos had to be replaced by 18-ton tractors. Often enough it took two of them to move the individual gun loads.

Left: Members of the unit examine shot-down Russian KW-1 B tanks at Gauvi on the road to Luga. Their crews, in the course of desperate attacks, had tried to interrupt the German advance on the road and had thus come within range of the 88 mm Flak guns.

A new firing position is reached. The crew of an 18-ton tractor tries to move the gun mount into the prepared firing position. After most of the artillery generator autos had broken down, this became quite complicated.

Over short distances, the 18-ton tractors could even tow two gun loads, here the barrel and mount wagons, photographed in August 1941.

Right: On August 10, 1941 the II./ 84 moved their guns into positions before Luga. In the process, the left-half bed wagon broke through the corduroy road.

Finally the job is done! The gun loads were towed laboriously to the prepared bed ditch and the two halves of the gun bed could be attached together and lowered.

Above: The gun is ready to fire.

Below: The first shot is fired a short time later.

In the latter half of September 1941, the six heavy 24 cm cannons of the II./84 were moved into positions very near the suburbs of the metropolis of Leningrad. In order to be protected from the fire of Soviet naval artillery, the gun crews surrounded their positions with walls of earth.

Below: An exciting photo by an amateur. While the dust stirred up by the shot slowly sinks to the ground, the aiming gunner runs to the gun with his aiming tool in his arm. The aiming tool was removed during every shot because of the severe shaking.

Right: The range of the heavy 24 cm cannons from their firing positions before Leningrad.

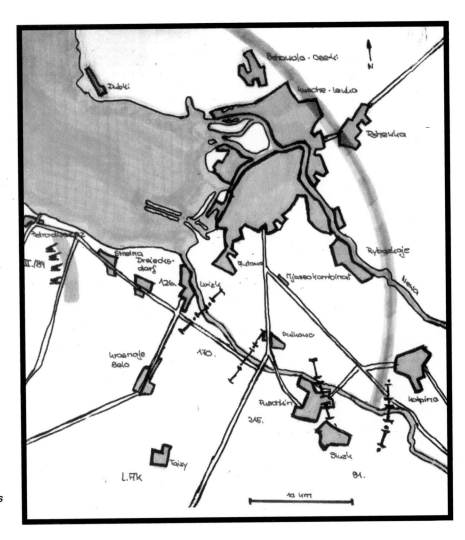

Below: A look at Leningrad from one of the II./84's observation posts.

Left: Typical army life: operating the guns. A gunner carries the still-smoking shell cases away (empty weight about 21 kg); his comrades prepare for the next shot. At the left, the aiming gunner, who has reattached the aiming tool, can be seen.

Below: Respectfully, the gunners observe the effects of Russian naval artillery shells, which left considerable craters in the meadows not far from the gun positions.

Right: This too was part of army life: a picture of the "fat grumbler" and its crew for a photo album.

Below: Before the action starts, the shells lie on wooden racks, cleaned and greased. The carrying tongs are attached to the shell in the middle; with their help, the shells were transported to the loading tray.

As with its action in France, the service of the II. Unit of Army Artillery Regiment 84 in Russia was often repoted on in the press, the radio and the newsreels. In No. 22 of the magazine "Die Wehrmacht" of October 22, 1941, an extensive picture story appeared.

Schwere Artillerie im Raum von Leningrad. Seit den frühen Morgenstunden des sonnigen Herbsttages speien unsere Batterien Tod und Verderben in die feindlichen Stellungen. Soeben hat eine Granate das Rohr verlassen und die Bedienungsmannschaft bringt das Rohr zum neuerlichen Laden in die Waagerechte. Die Kanoniere haben in der „Hitze des Gefechts" den Oberkörper entblößt, um bei der Geschützbedienung möglichst unbehindert hantieren zu können

Der schwere Brocken ist herangeschleppt. Fünf Mann werden ihn im nächsten Augenblick in das Rohr stoßen. Der Verschluß fliegt zu. Das Geschütz wird durch den Richtkanonier auf die errechnete Erhöhung eingestellt. Alles ist bereit und wartet auf den Feuerbefehl!

Feuer! Die Abzugsleine ist gezogen. Mit gewaltigem Donnerschlag jagt die Granate aus dem Rohr in das feindliche Ziel. Die Geschützbedienung läuft zurück, um den nächsten Abschuß vorzubereiten
PK-Aufnahmen: Kriegsberichter Günter Thiede (Wb 4), Kriegsberichter Spitta (1)

Treffer! Auf einem bei Kronstadt auf Strand gesetztem sowjetischen Schlachtschiff steigt eine riesige Qualmwolke empor. Die Schiffsgeschütze des Feindes werden so zum

6

The motorized artillery unit had an extensive array of various motor vehicles. This picture was taken in September 1941. Next to the 4th battery's small Horch Type 40 radio truck (Kfz. 17) stands a light, off-road-capable Stoewer Type 40 Uniform PKW (Kfz. 1). The following plates have been attached:
1. right fender (seen from front), the troop symbol (wolf's head),
2. left fender, the tactical symbol.

Below: October 1941, an electric chassis with a gun load in the mud.

The bottomless mud was soon followed by the first snow. Here an 18-ton tractor pulls one of the few still-intact artillery generator autos with an empty electric chassis and a second chassis with the gun mount toward the next railroad station.

Below: An artillery generator auto and bed load before being sent back to Germany.

Left: In October 1941 the first snow fell. The freezing gunners stand around the gun in firing pauses. The barrels were wrapped to decrease cooling.

Center: In a few days the temperature sank farther. Now the guns were completely covered and kept warm during long firing pauses. The purpose of this was to slow barrel cooling so that when orders came to fire, the gun could operate normally as soon as possible.

To fool the Soviet artillery, dummies were made of ammunition boxes, sheet metal and cardboard.

In the winter of 1941-42, the Army Group North gathered more flat-fire guns before Leningrad. Not far from the II./84's positions, these 28 cm cannons on railroad mounts took up positions.

Below: In 1942 the II./84, in its positions before Leningrad, was much weakened by equipment breakdowns. Only in the spring did the 4th battery receive three captured French guns to replace their two 24 cm cannons. In terms of their design, the 19.4 cm GPF (f) cannons on Schneider 485 (f) self-propelled mounts were similar to the C-trains of the heavy 24 cm cannons. Here too, the foremost vehicle of the train created the electric power to move the rest of the train. The generator and gun vehicles were separated when in firing positions.

Left: The maximum range of the 19.4 cm cannon was some 16,500 meters; the steel Type 21 (f) shell weighed 83.5 kg and was fired with a muzzle velocity of 725 meters per second. Compared with the heavy 24 cm cannon, this weakened the 4th battery's firepower. The greater mobility of the guns in the area of action offered some benefit in exchange.

Below: The 4th Battery in action. Technically, the 19.4 cm cannon was no longer satisfactory in 1942.

The 6th Battery of the artillery unit also returned to the Army Group North in the spring of 1942. In place of its 24 cm cannons, it had received three 17 cm cannons in mortar mounts. With the 68-kilogram 17 cm 39 shell, a maximum range of 28,000 meters could be attained; the muzzle velocity was 860 meters per second.

Below: The 5th Battery of the II./84 retained its heavy 24 cm cannons. Early in 1944 all the beds and mounts were lost to an offensive of the Red Army. The Russian troops, advancing stormily, also got their hands on one of the originally rescued barrel vehicles during the German retreat.

Technical Data

Crew:	16 men
Caliber:	240 mm
Muzzle velocity:	M.16 shell: 750 m/s
35(t) shell:	794 m/s
40(t) shell:	799 m/s
Rate of fire:	1 shot/min—1 shot/4 min.
Shot range:	M.16 shell: 26,300 m
40(t) shell:	29,875 m
Range of elevation:	-5 to +41.5 degrees
	(as heavy 24 cn Cannon (t))
	+10 to +42.5 degrees
	(as 24 cm Cannon M.16)
Loading position of barrel:	+6 degrees
Range of traverse:	360 degrees
Barrel length:	9600 mm = L/40
Barrel life:	1000 shots
Height of fire:	1850 mm
Recoil length:	1150 mm
Dimensions of bed ditch:	
Length:	6500 mm
Width:	5200 mm
Depth:	1400 mm

Weights:	
Gun in firing position:	79,100-86,000 kg
Barrel weight:	20,300 kg
Mount weight:	7400 kg
Cradle with gearing:	8000 kg
Bed vehicles ready to march:	
Left half:	36,600 kg
Right half:	37,600 kg
Mount vehicle:	30,800 kg
Barrel vehicle:	38,000 kg
Time of installation:	6-28 hours
Time of dismantling:	2.5-6 hours
Marching speeds:	
On road, rubber tires:	16 kph
On road, iron wheels:	12 kph
Prescribed marching speed:	10 kph
Rail speed:	27 kph
Climbing ability:	
On road:	26-29%
On rail:	9%

Above: The aiming gunner using the aiming tool.

Below: Artillery Generator Auto M.16 with bed load.

Ammunition Data

Loaded separately, shell case 240 x 1040 mm, empty weight 20.95 kg, two charges

1. Small charge: 45.15 kg Ngl.R.P.M.38 (14 x 4.3970)+ 2nd charge of 0.5 kg black powder
2. Large charge: 45.15 kg Ngl.R.P.M.38 (14 x 4.2/970)+ 11.3 kg Ngl.R.P. M.38 (14 x 4.3/970)+ 2nd charge of 0.5 kg black powder

Shells:

24 cm M.16 shell: Weight 215 kg, no further data

24 cm 21(t) shell: Weight 185 kg with hood, impact igniter 21(t), no further data

24 cm impact igniter shell 35(t): no data

24 cm impact igniter shell 35(t), revised: Weight 198 kg, explosive charge 24.663 kg, impact igniter CHZR(t) or SKHZR(t)

24 cm 40(t) shell: Weight 198 kg, explosive charge 23.371 kg, double igniter S/90 St. or impact igniter AZ 23 v. (0.15) and ground igniter DVZR(t)

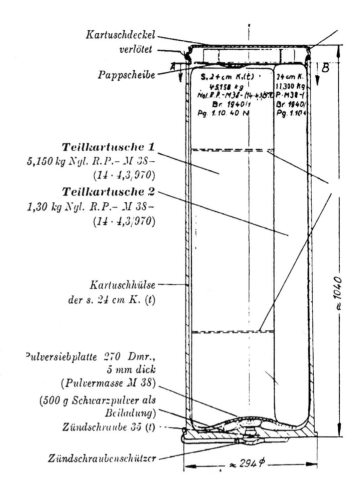

Above: Case Shell 35

24 cm impact igniter shell 35, revised in ammunition dept. Four ammunition gunners have applied the shell tongs to carry the 198 kilogram shell to the loading tray.

24 cm Impact Igniter Shell 35 (t)

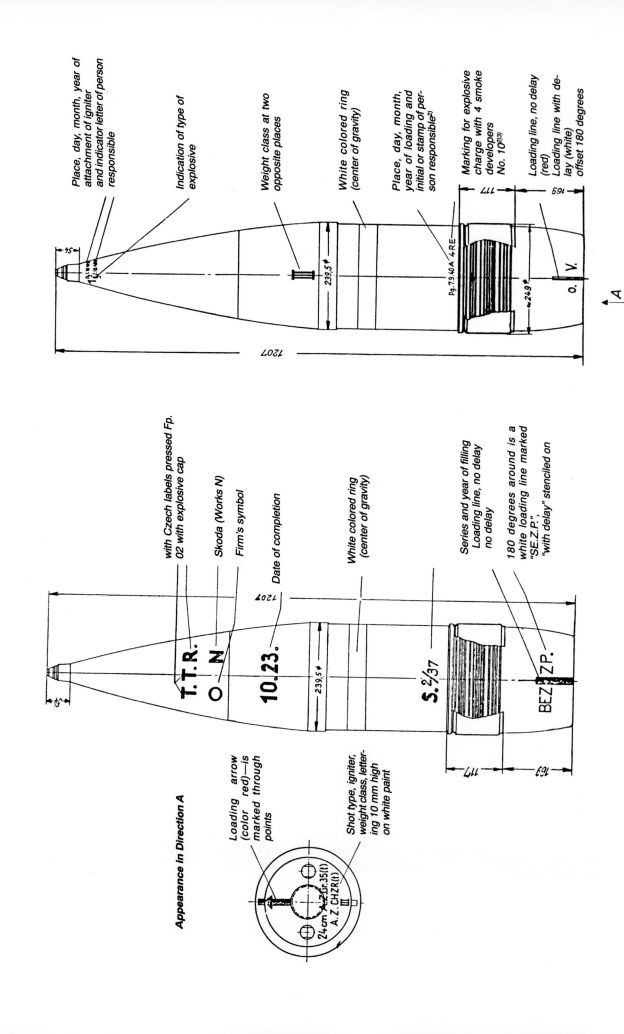

Place, day, month, year of attachment of igniter and indicator letter of person responsible

Indication of type of explosive

Weight class at two opposite places

White colored ring (center of gravity)

Place, day, month, year of loading and initial or stamp of person responsible[2]

Marking for explosive charge with 4 smoke developers No. 10[2)3)]

Loading line, no delay (red)

Loading line with delay (white) offset 180 degrees

with Czech labels pressed Fp. 02 with explosive cap

Skoda (Works N) Firm's symbol

Date of completion

White colored ring (center of gravity)

Series and year of filling Loading line, no delay no delay

180 degrees around is a white loading line marked "SE.Z.P."; "with delay" stenciled on

T.T.R.
O N
10.23.
S. 2/37
BEZ ZP.

Appearance In Direction A

Loading arrow (color red)—is marked through points

Shot type, igniter, weight class, lettering 10 mm high on white paint

24 cm A. Z. Gr.35(t)
A. Z. CHZR II

46

24 cm Shell 40 (t)

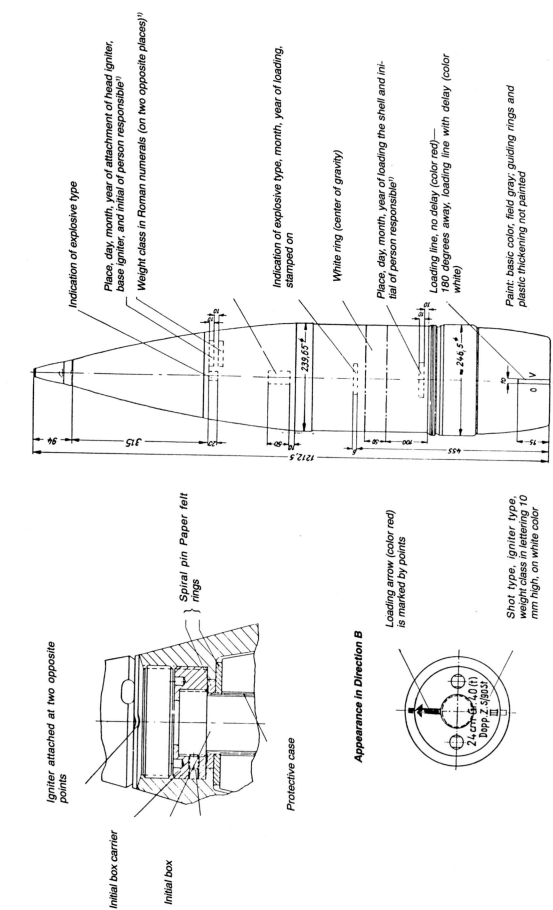

Indication of explosive type

Place, day, month, year of attachment of head igniter, base igniter, and initial of person responsible[1]

Weight class in Roman numerals (on two opposite places)[1]

Indication of explosive type, month, year of loading, stamped on

White ring (center of gravity)

Place, day, month, year of loading the shell and initial of person responsible[1]

Loading line, no delay (color red)— 180 degrees away, loading line with delay (color white)

Paint: basic color, field gray; guiding rings and plastic thickening not painted

Point "A"

Igniter attached at two opposite points

Initial box carrier

Initial box

Spiral pin Paper felt rings

Protective case

Appearance in Direction B

Loading arrow (color red) is marked by points

Shot type, igniter type, weight class in lettering 10 mm high, on white color

Military History Museum of Dresden

The largest special museum of its kind in Germany presents, on more than 7000 square meters of surface, more than 6000 items from the late Middle Ages to the present. Among them are many objects of international rank, such as the medieval staff-ring gun "Lazy Maid", the first German submarine "Brandtaucher", the landing gear of the Soviet spacecraft "Solyuz 29".

Offerings: Two exhibits at the Fortress of Königstein, interesting special and guest displays, manifold program for education and leisure.
Conferences (to 150 persons) possible.
Dining on the premises.

Hours: Tuesday to Saturday, 9:00 AM to 5:00 PM
Closed Mondays

Address: Olbrichtsplatz 3, 0-8060 Dresden, Germany

Telephone: 0037-51-592 3250

A look at the exhibition halls of the Military History Museum of Dresden. Above: a 4.8-ton explosive shell of the 80 cm railroad cannon "Dora". Below: Tank Destroyer 38 "Hetzer", 7.5 cm and 3,7 cm socket antitank guns aimed at the front of Soviet tanks.